Table of Contents

Science Action Labs

CI10139 Copyright © Teaching & Learning Company, Carthage, IL 62321-0010

Dear Teacher or Parent,

The spirit of Sir Isaac Newton will be with you and your students in this book. Newton loved science, math and experimenting. He explained the laws of gravity. He demonstrated the nature of light. He discovered how planets stay in orbit around our sun.

These labs are the product of many years of classroom science teaching. They have been used and revised based upon the hands-on experience of a wide variety of students. Pick and choose those that fit your personal approach to science teaching. Add, delete or modify to fit your style and your students' needs. Nothing I have written is engraved in stone. Place your personal imprint on them. Enjoy them as you enrich your students' lab activities.

The labs in this book are designed to be used directly with students. The background material and instructions are written for them. The assumption is that they are organized and graded as lab teams. Simply copy what you need for each classroom. Note that **Materials Needed** and **For the Teacher** are at the end of some labs. Delete those parts when copying for students to use.

Every teacher guide is supposed to have a basic philosophy. Mine is expressed in the goals listed below. They are based upon many years of meeting students' hands-on science needs.

Goals of a Laboratory Lesson

- To involve students in their science education.
- To get the teacher out of the dominant role and into the role of pot stirrer.
- To provide opportunities for creative, process-oriented thinking.
- To enable students to hypothesize and test their hypotheses within the limits of their background and laboratory materials.
- To provide opportunities for handling science apparatus.
- To let students enjoy sciencing.
- To provide opportunities for interaction of ideas within a group.
- To open new horizons in science education for the students.
- To confront students with some of the "unsolvable" aspects of science.
- To encourage the synthesis of their background facts and concepts in application to new problems.
- To provide "barriers" to make students conscious of their own limitations and thus encourage independent or classroom pursuit of science studies.
- To develop an appreciation for the fact that science can progress through a series of failures.
- To encourage careful observation and measurement.
- To accentuate the partnership of math and science.
- To encourage the big idea—the large conceptual pattern rather than the minute and isolated fact.
- To provide opportunities to summarize findings and to try to bring order out of apparent chaos.
- To encourage students to apply the same thought processes to other areas of their lives.

Sincerely,

Ed

Edward Shevick

SCIENCING

Active Hands-On Labs on the Scientific Method

Written by Edward Shevick
Illustrated by Marguerite Jones

Teaching & Learning Company

1204 Buchanan St., P.O. Box 10
Carthage, IL 62321-0010

This book belongs to

_____

The activity portrayed on the front cover is described on pages 45-46.

Cover design by Kelly Bollin

Copyright © 1998, Teaching & Learning Company

ISBN No. 1-57310-139-7

Printing No. 987654321

Teaching & Learning Company
1204 Buchanan St., P.O. Box 10
Carthage, IL 62321-0010

Philosophy for a Successful Science Program

In our education as science teachers, we have been repeatedly exposed to long lists of steps in the scientific method. In educating our own science students, we have also stressed the scientific method. Somehow, in all those process lists, we have forgotten that the scientist himself or herself is central to the scientific method. His or her intelligence, curiosity, scholastic skills, rational skills, humor and personality all play a part in creative science.

The central person in any science program is the youngster who could become a scientist or at least scientifically literate. Studies show that most beginning high school students have lost their motivation and ability to question and be curious. Contributing to this motivation loss are teachers and curricula that fail to recognize the characteristics and needs of these youngsters who have so much to offer society. The nature and beauty of science education is that it is capable of rekindling the intellectuality and curiosity of youngsters. Following are some elements in a successful science program:

Variety. Provide a variety of outlets for their intellectual curiosity. Plan for a variety of learning experiences both within the day and the semester. Minimize rote drills.

Time. Allow students sufficient time to generate ideas and incubate concepts. Full development of ideas may require periods of independent or isolated thinking.

Skills. Develop skills in reading, typing, library use, computer use and process science.

Interaction. Ample opportunity should be given for interaction between students, teachers and supportive adults. They should develop a sensitivity to others and an acceptance of non-conformity.

Evaluation. Let the students share in assessing their strengths and weaknesses. Don't assign arbitrary and old-fashioned standards to youngsters.

Creativity. Encourage creativity by being creative yourself. Be willing to "risk" a new kind of experience. You'll soon find your creativity mirrored in your students' activity as they try new approaches without fear of failure.

Independence. Children need a maximum of freedom and autonomy. Provide opportunities for self-directed discoveries and divergent thinking.

Acceptance. Try to accept their ideas even though they appear to be unorthodox. Accept their positions and feelings. Prepare them to accept some failure in their experiments.

Honesty. Demand the highest level of intellectual and personal honesty in their experiments and reports. Always expect their best work.

Maturity. Students are still children with fears, egos and problems. They may need your help in adjusting to your classroom environment.

How to Have a Successful Lab
For the Student

Creativity. You are encouraged to be original and ingenious in carrying out the laboratory assignments. Do new things or find new ways to do old things. Be creative in gathering materials for the laboratory, in handling the materials and especially in making reports to your teacher or your class.

Initiative. You will be expected to do as much as you possibly can on your own. Study your text for background knowledge and then use the concepts you have studied to solve the problem at hand. If you definitely need help, ask your teacher, but do not expect him or her to give you all the answers. He or she will merely spin you in the right direction and leave the rest to you.

Concepts. As you do the laboratory investigations, use the major concepts you have studied in the text and in class to solve the specific problems you find. Concepts are the big ideas of science.

Math and Accuracy. Many of the investigations involve measurement, data gathering, organizing and graphing. You will find that these methods often provide information that cannot be obtained in any other way. To be a competent science student, therefore, you need to be an accurate mathematician.

Reports. Write your laboratory reports in such a way that they are neat, accurate and complete. Fill in all data, question and conclusion sections.

Materials. Simple materials are called for in many investigations. Help the class progress by bringing in as much as you can.

Safety. Think safety. If you have the slightest doubt about safe procedure, check with your teacher. Do not endanger yourself, your classmates or your equipment.

Cleanup. You and your teammates will be expected to keep your laboratory area clean and cooperate with your teacher in cleaning up the room when the laboratory period is over.

Teams. In most of the laboratories you will working with teams of four. By doing this, you can help one another with the investigations just as professional scientists cooperate with one another to coordinate their activities.

Name _____

Observation Lab

Careful Observations

Scientists must be very careful how they observe things. They may use any of their five senses when they observe. **Observation** to a scientist means looking at a flower's color or listening to a bird's song. It may mean smelling a polluted pond or touching a furry animal.

Observing is not always easy. What you observe often depends on what you are trained to observe. The average person may observe the beauty in a butterfly's wing. An entomologist might observe how the wing shape differs from other kinds of butterflies. An ornithologist would note which birds catch and eat the butterfly. A meteorologist might observe how the wind affects its flight.

The scientific method depends upon careful, accurate observations. This investigation will help you practice and improve your observation skills.

You Can Fool Some of the People

Take a good look at this optical illusion. If you think your eyes are fooling you, you are right. Your eyes, as well as your other four senses, can be fooled easily.

Have each teammate make the best copy of this optical illusion on *their own paper*. Staple the copies to this lab when turned in for bonus points. *No tracing and no rulers.*

Optical illusions are constructed to cause your eye to make faulty observations. The designers of cars, public buildings and even clothing use optical illusions to affect how you observe things. Your eye is your most important observation sense. Yet it can easily be fooled. Scientists must be very careful that their observations are accurate.

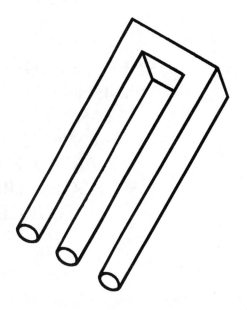

Name _____

Observation Test 1

What is wrong
with these scissors?

Observation Test 2

What is wrong
with this sign?

PARIS

IN THE

THE SPRING

Observation Test 3

Read the sentence
below and count how
many Fs there are.

**FIFTY FRIENDLY FARMERS
FINISHED FIFTEEN HARD
YEARS OF SCIENTIFIC
FARMING OF FIELDS
PLANTED IN LOADS
OF SOFT PERFUMED
FLOWERS.**

Observation Test 4

What is wrong
with the inscription on
this gravestone?

In Memory of
HERMAN JOHNSON
Born June 20, 1751
Died December 21, 1826
Age 75
In Memory Also
of His Widow
HILDA JOHNSON
Born August 29, 1755
Died October 24, 1825
Age 70

Name _____

Feeling Good

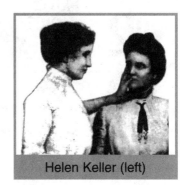

Helen Keller (left)

Helen Keller was a famous American woman. When she was two, an illness left her both blind and deaf. She observed the world around her by her sense of touch. She learned to read **braille** which is a system using dots on paper. By touching lips, she was able to understand what people were saying.

Helen Keller improved her sense of touch with practice. Can you improve your sense of touch with practice?

Your team will check each other's ability to recognize cut-out alphabet letters by touch only. Follow this procedure:

• Sit on a chair. Your eyes should be closed and your hands should be behind your back.

• Your team will hand you eight cut-out letters or numbers one at a time. Your job is to observe and identify them using *only* your sense of touch.

• Record the number of letters each student identified correctly in the Touch Data Table.

• Repeat for each member of your team.

TOUCH DATA TABLE			
1st Student	**2nd Student**	**3rd Student**	**4th Student**

Why do you think blind people have better than average senses of touch and hearing?

_____

_____

Observation Freedom

Plan an experiment in which you could test your teammates' senses of touch, taste or smell. If time permits, your teacher may allow you to carry out your experiment.

Materials Needed: You will need 50 to 60 cardboard or plastic letters or numbers at least 1¹/₂" (4 cm) high.

For the Teacher: Consider Observation Freedom optional. Stress safety and sanitation on taste and smell tests.

Name _____

Communication Lab

What Do I Do Next?

When American astronauts land back on Earth, they have traveled hundreds of thousands of miles (kilometers). Yet they land exactly on their Florida or California airport.

Part of the credit for their successful mission goes to their skill in carrying out very exact orders. A slight mistake could mean disaster because they could miss the Earth and drift off into space.

Your success as a science student depends upon improving your *communication* skills. Communication in science involves:

1. Being able to understand and follow instructions.

2. It also means being able to report clearly on the results of your experiments.

Let's practice communication.

Astronaut Test

Can you imagine finding this creature on the planet Mars? To help you learn to observe and communicate, here is a simple test.

Name _____

Do *only* what you are told to do. When you have finished, turn this paper over and wait quietly for your teacher to communicate with you.

Step 1   Don't do anything until you have read all 10 steps in these instructions.

Step 2   Print your name on the animal's body.

Step 3   Draw an *X* on each leg.

Step 4   Draw a □ around each eye.

Step 5   Draw 10 dots in the tail. (• • • • • • • • • •)

Step 6   Draw teeth in the mouth.

Step 7   Draw a line from the top of one ear to the other.

Step 8   Place 10 small circles inside the neck. (○○○○○○○○○○)

Step 9   Write *I am following orders* under the face.

Step 10   Now that you have finished reading these instructions, do only steps 1 through 4. Turn this paper over and sit quietly until your teacher communicates.

Scientists Don't Use Gobbledegook

Part of a scientist's job is to communicate with others. As a science student, you should also be able to discuss or write your experiments so that they are easily understood.

Writing that uses big words instead of little ones and wanders in circles is called **gobbledegook**. Gobbledegook leaves the reader confused.

On the following page is a list of common sayings that have been twisted into almost meaningless gobbledegook. Try to figure out what they mean. Don't expect any help from your teacher. He or she is more confused than you are.

Example: All elements that sparkle brilliantly don't have to be bullion.
Translation: All that glitters is not gold.

Name _____

Common Sayings in Gobbledegook

Read the five gobbledegooks below. Write what you think each gobbledegook is trying to say.

Gobbledegook 1:   The prompt, warm-blooded flying animal can apprehend the slender, crawling earthbound creature. _____

_____

Gobbledegook 2:   The edible red or green circular fruit can restrict the visitation of the medical profession. _____

_____

Gobbledegook 3:   Refrain from inspecting the present of a beast of burden in the place where food enters. _____

_____

Gobbledegook 4:   Do not compute feathery fowl previous to their emergence from their thin shell abode. _____

_____

Gobbledegook 5:   The rodents will participate in festive activities if the stealthy feline is not present. _____

_____

Gobbledegook Bonus

Try writing a gobbledegook of your own using an old saying, a science-related saying or your favorite expression. Try it on your classmates.

_____

_____

Making Scientific Decisions

The Scientific Method

The lab activities in this book have encouraged your use of the scientific method. The scientific method involves observing, questioning, measuring, experimenting and verifying your experiments. The scientific method is more than a set of standard procedures. To be successful, it must be blended with imagination and creative insight.

You can use the scientific method in everyday life. You make intelligent decisions on what foods to eat, what items to buy and how to spend your time. Even packing for a vacation calls for careful scientific decisions.

Scientists planning to explore the oceans or space have important scientific decisions to make. Their decisions can mean life or death for aquanauts and astronauts.

You Are Going on a Trip

To the Moon

You have been chosen for a trip to the moon. You can only bring 10 items with you. List the 10 important items you will need to survive.

1. _____
2. _____
3. _____
4. _____
5. _____

6. _____
7. _____
8. _____
9. _____
10. _____

Name _____

Into the Ocean Deep

Now you are going deep-sea diving. List the 10 most important items you should have with you.

1. _____   6. _____

2. _____   7. _____

3. _____   8. _____

4. _____   9. _____

5. _____   10. _____

A Scientific Vacation

You did well on your moon and ocean trip decisions. Here are some more scientific destinations. Pick *only one* trip and again list your 10 most needed items.

1. Long trip into an underground cave

2. Long balloon trip

3. Trip to the Arctic

4. Trip back 500 years in time

5. Trip ahead 500 years into the future

Have some science fun. Pick your favorite science trip and list six dumb things you might take along.

Your trip is to _____.

Your Dumb Items List

1. _____   4. _____

2. _____   5. _____

3. _____   6. _____

14

Science and Superstition

What Is Superstition?

Many years ago people had little science knowledge. They didn't know what caused disease, lightning or earthquakes. They developed superstitions to try to "explain" what they didn't understand.

Superstitions are mainly based upon ignorance and fear. Very few superstitions could stand up to a scientific test.

You've probably heard many superstitions. Don't take any of them too seriously. Ask yourself how they probably got started. Be skeptical. Don't believe a superstition just because your friends do.

Explaining Some Common Superstitions

Salt is common in many superstitions. This may be because salt was widely used to halt food decay. Throwing a pinch of salt over the left shoulder was supposed to keep evil away.

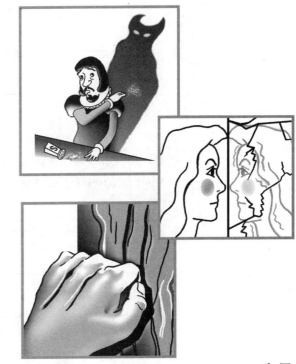

Superstitious people believed that breaking a mirror brought bad luck. They thought that the image in a mirror was part of their body. By breaking the mirror, they broke part of themselves.

Many people "knock on wood" when they discuss the possibility of something good happening. This superstition traces back to a belief in evil spirits. Loud knocking scared them away so they couldn't hear and steal your ideas.

Name _____

Spoofing Superstitions

On the following page is a list of common superstitions. Feel free to add to those superstitions by research or asking family and friends.

Pick out one superstition. Try to explain how it could have originated.

_____

_____

_____

_____

Make up a superstition of your own. Humor and imagination are encouraged. Embellish your superstitions with false facts and illogical arguments. Go down in history as the inventor of a superstition.

_____

_____

_____

_____

Pick a superstition that you can *safely* and *easily* test scientifically, for its truth or falsehood. Write your superstition plan below. Your teacher must approve your plan before testing the superstition.

_____

_____

_____

_____

Name _____

Superstition List

1. Cutting or shaving hair makes it grow faster.

2. Chubby people are better natured than slim people.

3. Dishonesty shows itself in a person's face.

4. Man has one less rib than a woman.

5. We "taste" food only by our tongue.

6. It is proved that people can know by mental telepathy what is happening at a distance.

7. Walking under a ladder will bring bad luck.

8. Handwriting is a clue to character.

9. Intelligent people are physically weaker than normal people.

10. Toads cause warts.

11. Health is determined by the stars under which you are born.

12. Exposure to the full moon can affect your mental health.

13. Older people hardly need any sleep.

14. Night air is unhealthy.

15. All athletes die prematurely.

16. Hair can turn white overnight from shock.

17. Lines and markings on your hand foretell your health and future.

18. People generally die about the same age their parents died.

19. Flowers are removed from hospital rooms at night because they compete with the patient for oxygen.

20. Fish is a brain food.

21. Birthmarks are caused by mothers receiving a shock before a child is born.

22. Thirteen is an unlucky number.

Scientists and Their Tools

Sciencing 5

NEWTON'S
ACTION LAB

Latin Scientists

Optometrists
use this chart.

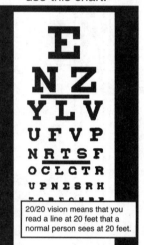

E
N Z
Y L V
U F V P
N R T S F
O C L C T R
U P N E S R H

20/20 vision means that you
read a line at 20 feet that a
normal person sees at 20 feet.

Plants and animals have Latin names. There is also a Latin name for each kind of science. Try to guess what each scientific field below studies. After your team guesses, your teacher will give you the correct answer. The *ology* in many science fields means "the study of."

Mystery Scientists: What Did They Study?

Hematologist: Charles Drew, American (1904-1950)
Ornithologist: James Audubon, American (1785-1851)
Anatomist: Andreas Vesalius, Belgian (1514-1564)

Science Field	Your Team Guess	Teacher Answer
1. Aeronautics		
2. Anatomy		
3. Anthropology		
4. Archaeology		
5. Dermatology		
6. Entomology		
7. Ethnology		
8. Botany		
9. Ornithology		
10. Toxicology		
11. Meteorology		
12. Paleontology		
13. Physiology		
14. Hematology		
15. Ichthyology		

Name _____

The Cool Tools of Science

Scientists need tools to observe and experiment. This activity is about some of the simpler scientific tools used in most laboratories.

On the following page are pictures of 12 science tools. Identify them and fill in their number on the Tool Identification Chart.

TOOL IDENTIFICATION CHART		
Name of Tool	**How Tool Is Used**	**Picture Number**
Example: Magnifier	To observe objects closely	9
Flasks	To handle and heat liquids	
Balance	To weigh things	
Thermometer	To measure heat	
Test Tube	To hold many experiments	
Microscope	To view very small objects	
Graduate	To measure liquids	
Petri Dish	To grow microbes in	
Tweezers	To handle small things	
Beaker	To hold and pour liquids	
Ring Stand and Clamp	To hold equipment for safe handling	
Burner	To heat things	

Name _____

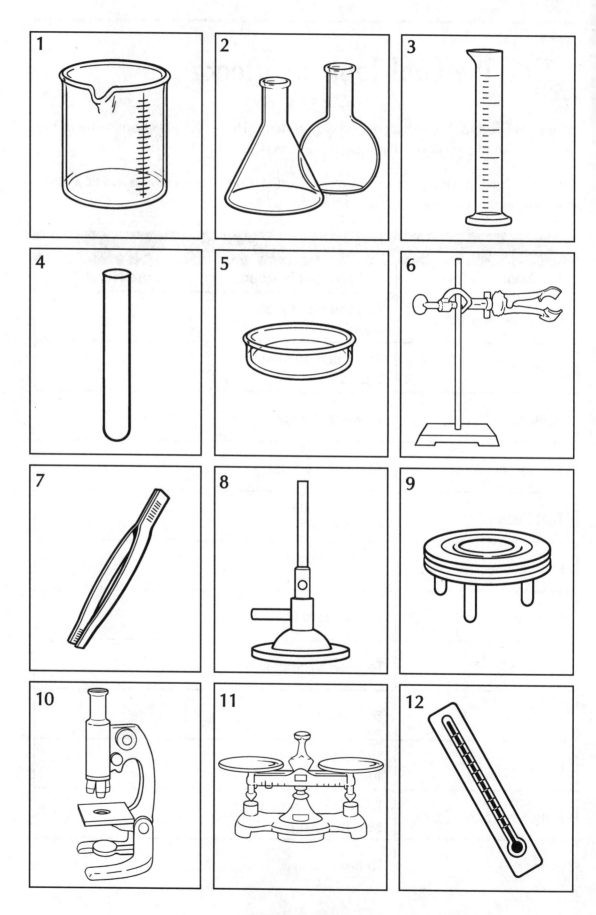

Name _____

Magnifier Fun

You are going to use a magnifier to observe many things. Here are some helpful hints.

In most cases it is best to view objects on a white piece of paper. Move your hand and adjust your head so maximum light falls upon the object viewed. Rotate the lens or move the entire magnifier back and forth for clearest view. Keep your fingers off the glass to prevent smudging.

You will be given many ideas as to what to observe with your magnifier. Identify your best five on the right and draw what you see in the circles provided.

Here are some viewing suggestions. Feel free to explore your own ideas.

sugar	salt
pepper	skin
stamps	coins
paper money	ink lines
fabrics	string
feathers	insects
leaves	

Materials Needed: Magnifiers and an assortment of odds and ends.

1. _____

2. _____

3. _____

4. _____

5. _____

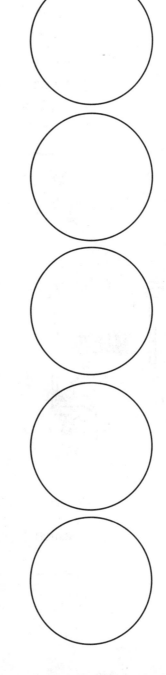

ACTION LAB

Sciencing

6

Hypothesis Lab

Observing, measuring, predicting and experimenting are all part of the scientific method. The scientific method of solving problems involves curiosity, creativity and imagination.

Predictions are possible solutions to a scientific problem. Scientists call their predictions **hypotheses**. Hypotheses are "educated guesses" that lead to experiments.

In this lab you will be given three scientific problems to solve. Your team is expected to discuss **all** three problems and answer questions 1 and 2 **before** doing any experiments with the equipment.

Study both the drawing and the actual equipment that you will be using. **Don't pick up the equipment.** Discuss each problem with your team and then answer questions 1, 2 and finally 3.

First Scientific Problem

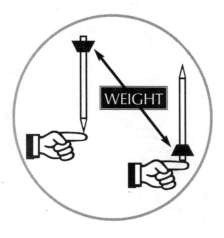

Your finger is used to balance a weight on a stick, as shown. Will the stick be easier to balance with the weight on the **top** or **bottom**?

Question 1: Your Prediction. (Which will balance best?)

_____

Question 2: Your Hypothesis. (Why do you think it will work as you've predicted?)

_____

_____

Now try to balance the stick.

Question 3: Your Test Results. (What really happened?)

_____

_____

22

TLC10139 Copyright © Teaching & Learning Company, Carthage, IL 62321-0010

Name _____

Second Scientific Problem

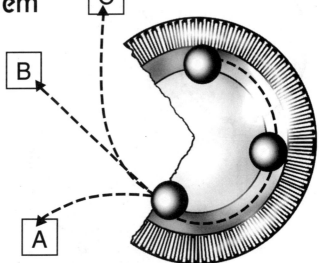

You will be given a pie pan with one-third cut out. Hold the pan down firmly on your table using one finger. Roll the ball rapidly along the rim so it exits at the open end.

What path will a rolling ball take after it leaves the cut metal pie pan? Will it follow path A, B or C?

Question 1: Your Prediction.

_____

_____

Question 2: Your Hypothesis.

_____

_____

Now try rolling the ball.

Question 3: Your Test Results.

_____

_____

Third Scientific Problem

One balloon is blown up to the size of a grapefruit. A second balloon is blown up to the size of a basketball. The two balloons are connected by one-hole rubber stoppers and glass tubing.

What do you predict will happen when air is allowed to flow freely from one balloon to another?

Name _____

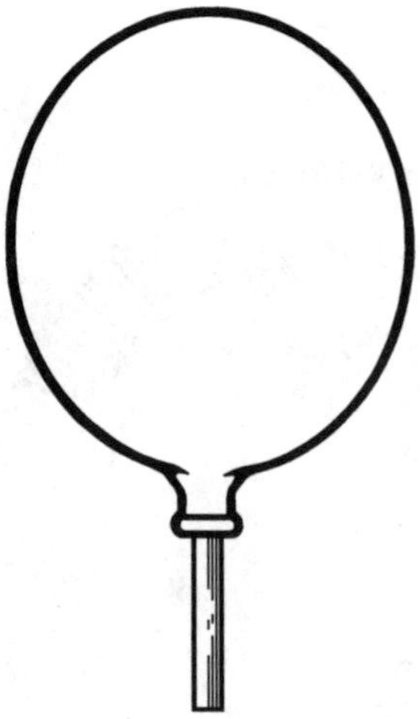

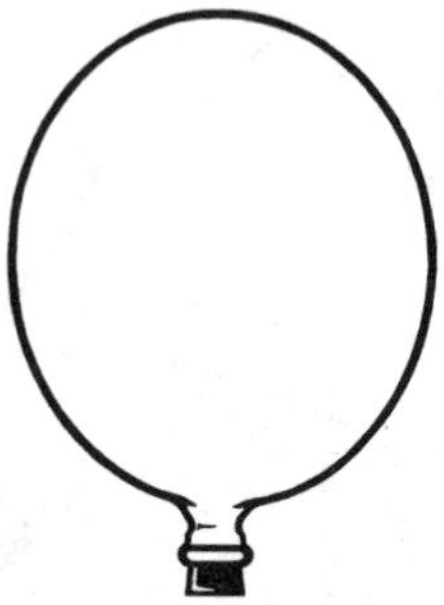

Question 1: Your Prediction.

_____

_____

Question 2: Your Hypothesis.

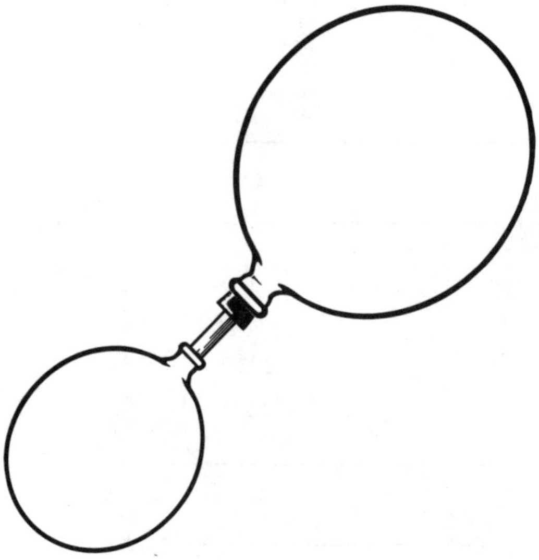

_____

_____

_____

Now blow up each balloon to the size required. Blow one up through the stopper and the other through the glass tubing. Pinch each balloon at the end while connecting them together. Unpinch and let the air flow where it will.

Question 3: Your Test Results.

_____

_____

_____

Materials Needed: You can make the weighted stick with pencils or doweling. Use the heaviest one-hole stoppers you have. It works best if you can construct the sticks out of metal. Make sure the stick extends at least 1" (2.5 cm) below the stopper.

Cut one-third out of an aluminum pie pan. Tape the rough edges for safety. Paper plates with deep rims will work. Use a large marble, ball bearing or golf ball.

Find one-hole stoppers that fit your balloons. The smaller end of the stoppers should attach to the glass tubing. Use about 5" (13 cm) of glass tubing and fire polish both ends for safety. It's best to use fresh balloons for each class.

Consumer Lab: Comparing Paper Towels

Can You Believe Advertising?

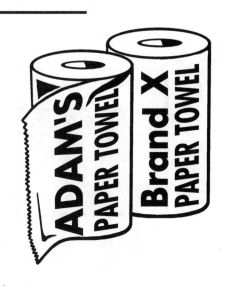

You are constantly being bombarded by ads for medicines, cosmetics, detergents and paper products. Each sponsor tells you how much better their product is compared to the competition. Some of the claims may be misleading or even false.

This lab will give you an opportunity to check claims made for various paper towels. You will be guided through two standard paper tests, and then you will be on your own.

Describe an ad you heard recently that seemed to have exaggerated claims.

_____

_____

_____

Absorbing Water

Absorbency is an important characteristic of paper towels. It is a measure of how much water a towel can take into its fibers. Let's compare the absorbency of three different kinds of paper towels.

1. Obtain a few sheets of three *different* kinds of paper towels. Mark each individual towel with its brand name.

Name _____

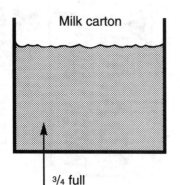

width →
5 cm
Brand X
8 cm line
Brand Y
8 cm
Brand Z
8 cm fold

PAPER TOWEL

Milk carton

¾ full

Small milk
carton   Strips

Brand X

2. Cut a 2" (5 cm) strip the full width of the sheet. **Notice:** Use the *width* and *not the length.*

3. Write the brand name in the center of each strip.

4. Mark a pen line *exactly* 3" (8 cm) from one end of each strip.

5. Fold all three brands on the 3" (8 cm) line.

6. Fill a small cut-down milk carton three-fourths full of water.

7. Place the carton of water in the center of your work table.

8. All three brands must be placed in the water at the **same time.**

9. Place the 3" (8 cm) folds over the edge of the carton so that they extend into the water. Place the other side of the three strips in *different* directions on the table.

10. Observe the absorption of water in each towel. Consider the race over when *one* strip is *completely covered* with water. Clean your area and answer the questions below.

Which brand of paper towel absorbed water the fastest?

_____

Which brand of paper towel absorbed water the slowest?

_____

Give some instances where speed of absorption is useful to the

consumer. _____

_____

_____

How Much Water Can They Hold?

A good paper towel should not only absorb liquids fast, it should also be able to absorb a great amount of water. This characteristic is called **saturation capacity**. Do some brands have a greater saturation capacity than others?

Consumer Lab: Comparing Paper Towels

Name _____

1. Obtain one more sheet of each of the same brands. Mark the brand name on each sheet.

2. Crumple and place each in a separate paper cup.

3. Weigh each cup and paper towel.

4. Record each weight in the data table under *Weight of Cup and Dry Towel*.

5. Place one brand of crumpled towel in the carton of water for **exactly one minute**.

6. Pull it out of the water and let it drip for one moment.

7. After one minute of dripping, place the towel in its **original** cup and weigh it.

8. Record the wet weight in the data table under *Weight of Cup and Saturated Towel*.

9. Subtract the dry weight from the saturated weight to find the total water absorbed. See the example for help. Record the answer under *Saturation Capacity*.

Repeat for your own two paper towels and record in the data table.

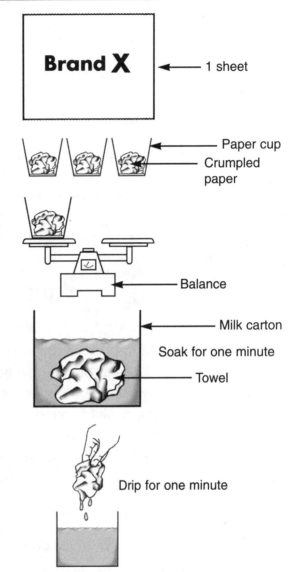

1 sheet

Paper cup
Crumpled paper

Balance

Milk carton
Soak for one minute
Towel

Drip for one minute

SATURATION CAPACITY DATA TABLE			
Brand of Paper Towel	**Weight of Cup and Dry Towel** Grams	**Weight of Cup and Saturated Towel** Grams	**Saturation Capacity** Grams
Example: Adam's Special	11.5	32.7	32.7 - 11.5 = 21.2
1.			
2.			
3.			

Name _____

Which brand of paper towel was saturated with the most water? _____

Which brand of paper towel held the least water? _____

Give some instances where saturation capacity is important to the consumer.

_____

_____

Paper Product Freedom

Absorption and saturation capacity are only two indicators of how good paper towels are. You are now going to have the opportunity to design and carry out your own tests on paper towels or other paper products. For example, you may test paper plates, paper cups, toilet papers, facial tissue, egg cartons, newspapers, writing papers, paper diapers or "?" **Important:** Your tests must compare three different brands of the same paper product. For example, you could test three different brands of facial tissue to find which is the best filter for polluted water.

Here is an outline for your paper product tests.

1. What do you plan to test? Give details plus diagrams.

2. What special materials will you need for your tests?

3. Design a data table to fit your tests.

4. What are your results and conclusions after testing your products?

Materials Needed: At least three different paper towel brands, paper cups, metric rulers, balances and cut-down small milk cartons.

For the Teacher: Supply the three brands of paper towels. For best results, use a premium brand, a cheap brand and any other type. The lab can get messy in terms of water spilling. Let students bring in the paper products for their own tests. Give plenty of planning time, but have the actual testing done in the classroom.

Mixed-Up Eggs

The Great Escape

Once upon a time there was a great scientist called Dr. Gregor Fendel. He set out to rediscover the laws of heredity. In his country they couldn't grow peas, so Dr. Fendel chose to experiment with chickens. After long years of crossing thousands of different types, he wound up with a very strange breed of chickens. These chickens were bred to lay only *hard-boiled eggs.*

Dr. Fendel absent-mindedly left his laboratory unlocked and all the "hard-boiled" chickens escaped. Eventually they spread throughout the world, and half the eggs now laid on a typical egg ranch are hard-boiled. That is why this science class has been hired by the Egg Growers Association to find fast, simple and sure methods of separating raw and hard-boiled eggs.

How to Proceed

1. You will be given two marked eggs. One will be raw and the other will be hard-boiled. Your teacher will know the code.

2. First **hypothesize** (make educated guesses) as to how you might test the eggs for differences.

3. Fill out column A on the Egg Data Table. Column A should clearly state *what* you plan to do.

4. Fill out column B on the Egg Data Table. In this column tell what differences you *expect* to find.

5. Don't experiment until *all* your hypotheses (both columns A and B) are signed by your teacher.

6. Your experiments must not harm or crack the eggs. *Demerit points* will be given to your team if the eggs are cracked. Explain your test results in column C.

Name _____

7. You may have a few hypotheses that are not testable with the materials available in our science class. Most of your hypotheses should be testable in class.

8. The purpose of this lab is *not* just to determine which egg is which. Emphasis is on devising and carrying out intelligent egg tests.

Egg Data Table

EGG DATA TABLE		
A. What Do You Plan to Do?	**B. What Differences Do You Expect?**	**C. What Happened? What Does It Prove?**
Example: Float both eggs in salt water.	Example: A boiled egg is lighter and should float higher.	Example: The egg marked __ floated higher. This proves that boiled eggs are lighter.
Teacher signature _____	Teacher signature _____	

Result: The egg marked _____ is hard-boiled.

Materials Needed: Two eggs are needed per team. One is raw and the other is hard-boiled. Mark them *X* or *Y* or any desired code.

You will also need standard science equipment such as balances, weight sets, beakers, lights, etc. Be ready for strange requests.

For the Teacher: Allow adequate time for students to complete columns A and B. Sign off on these columns before any testing is done. You might allow a full lab period for the actual hypothesis testing.

Name _____

NEWTON'S
ACTION LAB

Sciencing

9

Sciencing with Sound

Sound in Your Life

Close your eyes for a moment and concentrate on all the sounds that come pouring into your ears. Next to sight, the sense of sound is the most vital sense that links you to your environment. This lab will help explore the science of sound.

Your class will be divided into teams. Each team will plan and perform certain sound activities. No real musical instruments can be used. You are required to use straws, jars, boxes and other common equipment. Request your teacher's permission for any exceptions to the use of simple, inexpensive materials.

Battle of the Bands

Your teams will be given planning time to prepare to demonstrate certain sounds. You will judged excellent to poor on each of the sound categories below. Much of the work should be done at home.

TO DEMONSTRATE SCORE:	excellent 4	very good 3	good 2	average 1	poor 0
1. a pleasant sound					
2. an unpleasant sound					
3. a very high-pitched note					
4. a very low-pitched note					
5. a very original musical instrument					
6. the musical scale (do, re, mi . . .)					
7. a recognizable tune on one instrument (double points)					
8. all instruments played together for a band effect (double points)					

Total points scored by your team: _____

For the Teacher: Encourage library research on sound science. Your students will need three or four days between planning and the Battle of the Bands. They should plan in class and construct at home.

Name _____

Sciencing with Straws

Straws Are for More Than Drinking

The National Association of Straw Manufacturers is anxious to increase Americans' consumption of straws. In their search for ideas beyond sipping sodas and milk, they have turned to the fertile minds of students. They're offering around-the-world cruises (on a raft made of straws) to those who can best answer these questions:

- What novel uses can students dream up for straws?
- How can science principles be taught using straws instead of elaborate costly science equipment?

Materials You Can Use

- The prime ingredient of all your gadgets must be straws.
- You may use any other materials needed such as tape, pins, rubber bands, etc. All should be common and inexpensive. Most materials must be supplied by your teams. Get teacher permission for any exceptions.

Procedure to Follow

- Your team will be given planning and construction time in class. Some work may have to be done at home.
- Team reports will be presented orally to the class on _____.
- A short written report detailing and sketching your gadgets must be turned in at the same time.
- Plan to show two novel straw uses.
- Plan to show one science principle based on straws.
- Grading will be based upon: 25%–How original your novel straw uses are. 25%–How important and complicated your science principle is. 25%–Workmanship and clever uses of materials. 25%–How well you demonstrated your ideas to your class.

Materials Needed: Back up your students with extra straws, tape, pins and rubber bands.

For the Teacher: Have straws, tape, pins, etc., available beyond what your students may provide. You might get them started with a simple straw experiment. Dip one end of a straw in flour and blow it onto a candle or Bunsen burner flame. This shows rapid combustion. Provide adequate class planning and research time. Allow at least three days between the planning sessions and the team reports.

Observing a Candle

Candles

Candles have served people in many ways. Candles provide light and, in some cases, heat. Candles can decorate a room or set the mood at a candlelight dinner. Candles are used in many religious ceremonies.

Your body and a candle are alike in many ways. Both require food that is burned to give off heat. Both you and a candle use up the oxygen in the air and give off carbon dioxide gas.

Observation

In this lesson you are going to use your science skills to carefully **observe** a burning candle. A famous scientist named Faraday made so many candle observations that he wrote a book about candles.

Your teacher will light your team's candle. Observe it carefully. Write at least 13 things that you observe.

CANDLE OBSERVATION TABLE
1.
2.
3.
4.
5.
6.
7.
8.
9.
10.
11.
12.
13.

Name _____

You will now observe the candle in the dark. Can you observe six things that you hadn't observed before?

CANDLE OBSERVATION TABLE
1.
2.
3.
4.
5.
6.

Materials Needed: You can use regular candles, but birthday candles will do. Provide a very stable candle base.

For the Teacher: The teacher should light the candles for all teams. This is a good time to stress safety.

Invasion of the Insectpicks

Duplicating Darwin

Today you are going to play Charles Darwin. Charles Darwin was a famous biologist who carefully observed nature. He came to some conclusions about **evolution** that shook the world. His observations showed him that all species of animals undergo changes. When the changes hindered the animal in its environment, the changed species died out. When the changes helped the animal cope with its environment, the species prospered. For example, rabbits that developed a white fur in the snowy winter were relatively safe from their enemies. If rabbits had developed a red fur in the winter, they could easily have been spotted by their enemies.

What your class does today may never shake the world, but it might shake up the school gardener. You are going to be put into a situation from which Darwin would have drawn many scientific conclusions. The challenge is for you to come up with the same kind of conclusions after the experience.

Outside your classroom, there is a grassy area that has been **invaded** by insects. The "insects" will really be 150 multicolored toothpicks that have been broken in half and scattered at random on the grassy area. Let's call them **insectpicks**. There will be 30 red, 30 blue, 30 green, 30 natural and 30 yellow toothpicks.

You and your classmates will play the part of predator birds that just love to eat juicy **insectpicks**. The **insectpick** predators will be lined up surrounding the grassy area. When your teacher gives you the signal, your team will try to capture all the **insectpicks** you can find in four minutes. The teams collecting the most toothpicks will be given bonus points on the lab.

Name _____

Your team will fill out the Team Data Table back in the classroom. Each team captain will report the team data for the Class Data Table. All **insectpick** counts will be verified by the class lab assistant.

Question 1: Which color **insectpick** does your team believe will survive and prosper on the grassy lawn? Explain your answer. _____

_____

_____

 # Darwin on Your School Lawn

Now go outside and follow these instructions:

1. Line up outside surrounding the **insectpick** area.

2. Find as many **insectpicks** as you can in the time allotted.

3. Go back to the classroom and fill out the Team Data Table.

4. When called upon, combine your Team Data Table into the Class Data Table.

TEAM DATA TABLE	
Insectpick Color	Number of Insectpicks Captured
Red	
Blue	
Green	
Natural	
Yellow	

CLASS DATA TABLE			
Insectpick Color	Original Number	Number of Insectpicks Captured	Number of Insectpicks That Survived to Grow and Reproduce
Red	30		
Blue	30		
Green	30		
Natural	30		
Yellow	30		

Name _____

Question 2: From the Class Data Table, determine which color **insectpick** had the

highest survival rate. _____

_____

_____

Question 3: Name two animals and tell how their color protects them. _____

_____

_____

_____

Question 4: Write a paragraph telling the conclusions that you arrived at as a result of your observations of the grassy area and the class data on **insectpicks** that survived. Be brave enough to come up with conclusions that are valid for other animals besides birds and insects. Make Mr. Darwin proud of you!

_____

_____

_____

Question 5: The human species also undergoes changes. Use your imagination to come up with two possible human changes that might help man adjust to our modern environment. Tell how each change might help man.

A. _____

_____

B. _____

_____

Materials Needed: Provide multicolored toothpicks. Break them approximately in half. Provide containers for each team to hold their **insectpicks**.

For the Teacher: Use a different grassy area for each class. This is to insure that lost **insectpicks** will not affect succeeding class results. Place a copy of the Class Data Table on the board for the team summary. You may be surprised by which color is hardest to find. Have the colored toothpicks pre-packaged in groups of 30 for easy seeding of the grassy area.

Name _____

Mystery Box Lab: Learning by Indirect Evidence

Observing Without Seeing

Textbooks and teachers tell you that atoms cannot be seen. Yet scientists can tell you the differences in weight, position and electrical charges of particles found within the atom. If you are puzzled, it's because you are not aware of the fact that scientists can find evidence of a particle's existence without observing them directly with their five senses.

You can often see the vapor trail of a jet plane high in the sky without actually seeing the airplane itself. By studying the trail you could probably estimate the plane's height, speed and could spot collisions with other planes. In a similar way, atomic particles can leave a trail on special film or "cloud chambers" and can even be deflected by magnetism and electricity.

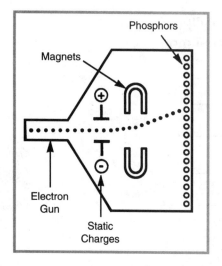

Your television picture is another example of indirect evidence. The picture is made by a stream of invisible electrons. The electrons are made to strike special chemicals (called phosphors) on the screen. No one has ever seen electrons, yet there they are being pulled around by magnetism and electric charges to show your favorite program.

In this lab you are going to see for yourself how objects sealed in boxes can be identified by the way they feel, move, weigh or sound. Scientists call this learning by "indirect evidence." How well you identify the objects in the boxes will depend on how many "questions" your team can ask the box. Try to do more than just shake the mystery boxes.

Preparing Mystery Boxes

Each team will prepare four mystery boxes. Label them with your team number and the letters *A, B, C* and *D*. Use small boxes such as bandage or individual cereal boxes.

TLC10139 Copyright © Teaching & Learning Company, Carthage, IL 62321-0010

Name _____

Place a different item in each box and seal it. Try to find interesting items that give some indirect clues as to what they are. Keep what's inside your boxes a secret from other teams. Bring them in on _____.

What's in the Boxes?

Your teacher will collect all boxes. They will be checked for proper identification and sealing. Ignore your own team's box as they are passed from team to team.

You are not expected to do all the boxes in one class period. The winning teams may be rewarded with bonus points on the next exam.

Mystery Box Scorecard

Describe *briefly* and *clearly* what you think each box contains.

Team 1	Team 2	Team 3
A _____	A _____	A _____
B _____	B _____	B _____
C _____	C _____	C _____
D _____	D _____	D _____

Team 4	Team 5	Team 6
A _____	A _____	A _____
B _____	B _____	B _____
C _____	C _____	C _____
D _____	D _____	D _____

Materials Needed: You will need tape to seal the boxes and markers to mark team number and box letter.

For the Teacher: Stop and start the box passing at your convenience. Keep the boxes moving around the room. Some teams tend to let the boxes pile up at their position. Give prizes or bonus points at your discretion. You could also reward the team with the most imaginative materials. Make a master list of what's in the boxes as each team submits them.

Name _____

Science Frontiers

The Greatest Science Discoveries

Look around your home and school. You can see dozens of devices from lights to computers that didn't exist years ago.

Here is a list of 30 inventions that have changed your life. Some are simple inventions and some are very complicated.

1. toilet	9. laser	17. bicycle	25. movies
2. computer	10. antiseptics	18. satellites	26. compass
3. printing press	11. plastics	19. refrigeration	27. thermometer
4. fire	12. airplane	20. money	28. glass
5. wheel	13. electric lights	21. telephone	29. zipper
6. radio	14. X rays	22. television	30. telescope
7. antibiotics	15. cars	23. photographs	
8. transistor	16. microscope	24. radar	

Which five of the above do you think are the most important? _____

_____

Which two do you think were invented earliest? _____

_____

Which two do you think were invented most recently? _____

_____

Which one invention do you think is the most helpful to people? _____

_____

Which inventions help us live longer? _____

_____

Name _____

Science in Your Present Life

Astronauts have walked on the moon. To do so, they had to develop rockets, special materials and electronic monitoring devices. The spin-off from space science gave us cordless power tools, solar cells and scratch-resistant plastic eyeglass lenses.

Study the Technology Chart below. It lists four technology fields and how they have led to inventions that have changed your life.

TECHNOLOGY CHART		
Technology	**What It Is**	**What It Can Do**
Lasers	Amplify light through emission of radiation	Measure speed and distance; cut metal, wood and diamonds; perform surgery; grocery bar codes
Geonomics	Study of genes and their place in DNA structure	Reverse genetic birth defects, create vaccines, clone new life-forms, slow aging
Biotechnology	Applies technology to the field of biological science	Develop tastier and more nutritious foods, create disease-resistant plants, create skin tissue and body parts, grow human organs in animals
Global Positioning	Uses satellites to observe the Earth and pinpoint positions on Earth	Navigate airplanes, ships, trucks and autos, pinpoint missing children or pets, improve farming by analyzing soil

Pick one of the technologies above. What do you consider its most important benefit to society? _____

_____

Pick a different technology above. Besides what is listed, what inventions would you like to see developed? _____

_____

Name _____

TIME TO CROSS THE ATLANTIC OCEAN

Year	Type of Transportation	Length of Trip
1850	Sailing Ship	11 Days
1936	Steamship	5 Days
1945	Propeller Plane	24 Hours
1958	Jet Plane	7 Hours
2000	Supersonic Plane	3½ Hours

Science in Your Future

Future scientists face many unsolved problems. How to feed a growing world population? How to keep certain plants and animals from becoming extinct? How to find if there is other life in our universe?

Some scientific discoveries can even be disturbing. Biologists in Scotland have learned how to **clone** sheep. They placed the cell of an adult sheep into the ovary of another sheep. The cell divided and developed into the world's first cloned animal. The scientists named the first cloned animal Dolly.

Debate time. The same technology that cloned Dolly can possibly be used to clone humans. You could make 100 copies of any person you choose. Is this a good or bad idea?

1. Form teams of three or four.

2. Describe whether you are for or against human cloning.

3. Research and develop your arguments for or against human cloning.

4. Your human cloning debate will take place on _____.

Name _____

Create a Creature Contest

Is Anyone Out There?

Ancient man imagined that many gods dwelled in the heavens. Modern astronomers still are uncertain as to whether life exists "out there." They are even less certain what form it would take if there were life.

Because of the recent scientific investigations, however, we can predict much more than ancient man could about the conditions to be found on various planets.

We don't expect to find life on other planets. But if there were life, how could it adapt to the conditions on each planet?

Your Planet Report

1. Research textbooks, the library and the internet to learn about planets.

2. Choose one planet on which to prepare a written report.

3. Your report should include data on the planet's size, day, year, moons, temperature, etc.

4. Your report should point out what is "different" about your planet.

5. Tell why you think life could or could not survive on your planet.

6. Spice up your report with sketches, graphs or photos.

7. Your written report is due on _____.

Name _____

Life on Your Planet

Assemble a creature with characteristics needed to survive on your planet. For example, if there is little gravity on your planet, its feet could have suction cups. If it is a cold planet, give your creature a fur coat.

The creature can be made of tin cans, light bulbs, plastic containers, Styrofoam™, marshmallows, milk cartons, doll parts, vegetables, old toys or anything in your garage or junk pile. You can even bake a cake creature.

Do not use any expensive parts. Your creature should be made of common, simple materials. You are not allowed to use a commercial kit.

Use your ingenuity and sense of humor to match your creature to the conditions on its planet. Attach a card to your contest entry giving your name, the creature's name and describing the special features.

Judging of contest will be done by class vote to select the best six. Your teacher will pick the top winners. Judging will be done on the following basis:

1. Clever use of materials to build the creature

2. Humor and imagination used to create the creature

3. Special features used to adapt the creature to its planet

4. Excellence of your class presentation

NEWTON'S
ACTION LAB
Sciencing
16

Canmobile Race

Converting Potential to Kinetic Energy

When you eat sugar, you are converting the **potential** chemical energy of the sugar into the energy we need for body warmth and motion. A gasoline engine converts the potential energy of the fuel into the **kinetic** energy of the moving car. All around you there are examples of how man converts **stored** potential energy into **moving** kinetic energy.

In this lab, you are asked to apply your scientific genius to building a better **canmobile**. This is a device for converting potential energy stored in a rubber band to moving kinetic energy. If you are successful, future generations will desert their Fords and Chevrolets to ride the freeways in your Canmobile.

Canmobile Race Rules

1. Your team will be given planning and testing time, but all Canmobiles must be built at home.

2. Your team must provide all materials except rubber bands. Use those provided.

3. The great Canmobile Race will be on _____.

4. You can only submit two entries per team.

5. For each entry you will be allowed two trials.

6. Each can must be marked with your team number.

7. The Canmobile must stay within the track. If your Canmobile leaves the track, that point will be considered its furthest distance.

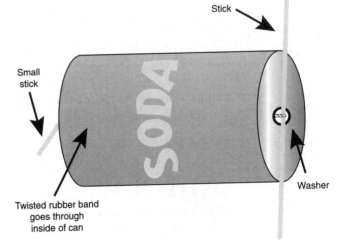

Stick

Small stick

Twisted rubber band goes through inside of can

Washer

Name _____

8. The Canmobile that goes the farthest within the track wins.

9. All Canmobiles must start out as **standard size** soda cans. What you do to modify the can is up to your inventive genius.

10. Helpful hints in building your canmobile:

- Use an empty **clean** soda can.
- Get help to punch a ¼" (.6 cm) hole in the center of the bottom of the can.
- Use a pencil or pen to push the rubber band through the inside of the can.
- Place a **short** stick through the rubber band at the bottom end. The stick can be taped to the bottom. It should not extend beyond the can.
- Place a washer through the other end of the rubber band.
- Place the long stick through the rubber band.
- Wind it up and win the race.

Canmobile Physics

Designing a winning Canmobile is an engineering challenge. Knowing the following physics principles will help you build a winning Canmobile.

Friction. When the moving parts of the Canmobile rub, they produce heat. This causes a loss of energy that could be used for forward motion. You can cut down friction by using a lubricant such as candle wax, oil or graphite. Try cutting down friction by using washers or by sanding or steel-wooling parts in contact.

Stability. Canmobiles have a tendency to spin off the track. You can increase stability by varying the size and length of the stick or using a stick on both sides. Adding weight to the can or using a nail or other metal can help.

Traction. Your can will tend to lose contact with the ground and spin so that it goes nowhere. This is the same as if your mom or dad drove around on smooth, bald tires. You can improve traction by roughing up the can rims, taping on rough paper or gritty adhesives. Adding weights to the can or stick will help traction as well as stability. Unfortunately, the added weight can increase friction and use up your stored energy.

Materials Needed: Provide four rubber bands to each team. All must be strong and exactly the same. You will also need soda cans, washers, sticks, candle wax.

For the Teacher: Your students will have lots of problems. That is the entire point of this lab. Let them think, experiment and sweat.

The track may be your room, the hallway or any appropriate indoor or outdoor area. Mark off at least a 6' (1.8 m) width. Any Canmobile that wanders off the track is flagged at that point. Have one team act to mark distances for all.

Name _____

Sciencing with Balloons

Balloon Science

The dictionary defines a *balloon* as "an inflatable rubber bag, impermeable to gas, often brightly colored and used as a child's toy." In science class, the balloon takes on another definition as a means of demonstrating science principles. In this lab, you will be led to discover some important concepts by "sciencing" with balloons. Your team will also be given the opportunity to demonstrate to the class an imaginative new use of balloons in any field of science from oceanography to space travel.

Note: To save balloons for use through all the experiments, don't tie the neck to seal the air in. Instead, twist the neck of the balloon and seal with a paper clip, masking tape or tie with thread or use a twist tie.

Balloon Experiments

Topography

Topographers study surface areas and relief features of the Earth in designing accurate maps. You are going to mark your balloon as shown in the diagram to the right to see how shapes change as a surface area is expanded. Use a pen to draw rectangles 1/2" (1.25 cm) long and 1/4" (.6 cm) wide every 3/4" (1.9 cm) on one side of the uninflated balloon. On the other side of the uninflated balloon, mark a small geometric design on the neck and top of the balloon. Now inflate the balloons to almost full size and draw the new shapes. Describe which areas of the expanded balloon gave the greatest shape distortions.

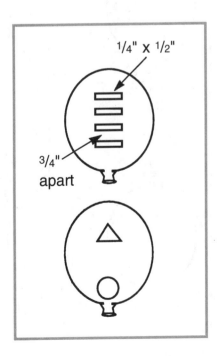

1/4" x 1/2"

3/4"
apart

47

Name _____

Static Electricity

Static electricity works best on dry days and depends on electric charges building up on nonconductors through friction. Balloons are non-conductors and by rubbing them with wool or silk or any garment you are wearing, you can build up a static charge on their surfaces. Try the following:

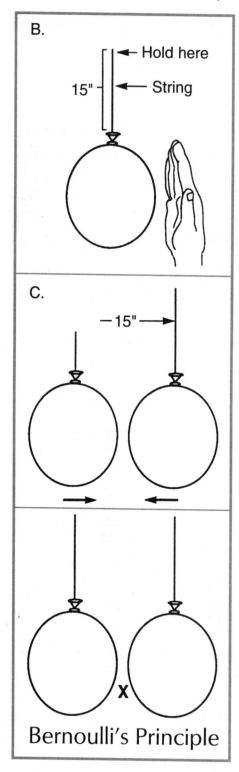

Bernoulli's Principle

a. Rub a fully inflated balloon vigorously and see if you can make it stick to an electrically neutral wall. (Don't seal with a paper clip for this experiment as the balloon becomes too heavy.) What does that tell you about the attraction of charged objects, such as the balloon, and neutral objects, such as the wall?

b. Tie 15" (38 cm) of thread on your balloon. Rub the balloon vigorously and have your teammates hold it as shown. Slowly bring your hand (a neutral object) toward it and see if you can attract the balloon toward you. Roughly estimate the distance at which your hand begins to attract the charged balloon. Try various ways of attracting the suspended balloon, and describe the results.

c. Tie two balloons as in part b and rub them vigorously. Hold them as shown and slowly bring them toward each other. What does that tell you about two objects that have the same type of charge?

Bernoulli's Principle

Bernoulli's principle tells us that when air is sped up, its pressure goes down. It can be used to explain why airplanes fly and baseballs curve. Use the same two balloons with 15" (38 cm) of thread attached as in the static experiments. Hold the balloons 1½" (4 cm) apart as shown and blow between them at the point marked X. Describe what happened and how Bernoulli's principle applies.

Name _____

Newton's Third Law

Rocket propulsion through space depends on Newton's law which states that for every action there is a reaction. When you fire a rifle, the bullet would represent the action and the rifle jerking backward the reaction. Blow up a balloon to almost maximum size and tape a 2" (5 cm) section of a large diameter straw inside the neck as shown to act as a rocket nozzle.

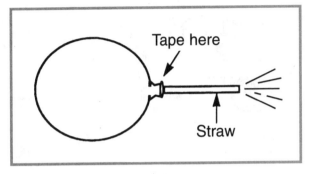

Hold your rocket engine high over your head and let it go. Describe what happened and tell in detail what you might do to improve the stability and thrust of your balloon "rocket."

_____

_____

_____

_____

_____

_____

_____

Balloon Freedom

1. Your team will be given time to plan an experiment involving a science principle with balloons.

2. Your demonstration in front of the class will be due on _____.

3. Feel free to use a balloon or part of a balloon in any (safe) way you wish to demonstrate a science principle. You can use it for insulation, as a slingshot, for storage or "?" Don't restrict yourself to blown-up balloons.

4. Your team grade will be determined by originality, your choice of science principle and success in class demonstration.

Materials Needed: You will need balloons; paper clips; string or thread; 2" (5 cm) straw sections; small pieces of wool, silk or fur and some markers.

For the Teacher: Allow research time for developing science principles with balloons. Insist that balloons be the basic component.

Sciencing with Air

Our Ocean of Air

14.7
lb.

Both people and fish exist in "oceans." People live in an ocean of air, whereas fish live in an ocean of water. Both get the oxygen they need from their ocean, and both pollute it with their waste products. Both are subjected to tremendous pressures due to the masses of air and water. How does this pressure of air affect your everyday life?

This lab contains the 14.7 pounds per square inch (1.07 km per sq. cm) of pressure that pushes down on all of us at sea level. This pressure makes possible the operation of many devices that most people take for granted. The air in your classroom probably weighs as much as you do.

Pipets: The Scientist's Straw

Scientists use pipets to transfer small amounts of liquid. Pipets are basically like straws. Your mouth decreases pressure near the top, and normal air pressure pushes the liquid up. Your task in this activity is to calibrate (measure) a pipet using air pressure to help you.

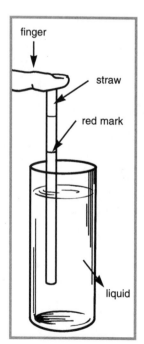

finger

straw

red mark

liquid

1. Obtain a pipet or straw marked with a red stripe, a 50-milliliter graduate, a clean cup of water and a cup to hold used water.

> **Caution!** For sanitary reasons, *only one member* of your team is to handle the pipet.

2. Use your mouth to draw liquid up the pipet slightly above the red mark.

3. *Quickly* place your finger over the top. Air pressure at the bottom of the pipet will keep the water in.

4. Remove your finger from the top of the pipet just long enough to run the water to the bottom of the red mark. This will take a little patience and practice.

Name _____

5. Transfer the fluid to the graduate.

6. Repeat the procedure four more times.

The graduate now has how many milliliters of fluid? _____ milliliters

7. Divide the total milliliters by five.

Your pipet is now calibrated so that each pipet holds how many milliliters up to the

red mark? _____ milliliters

Pipets are useful devices in many science laboratories.

Air Pressure Magic

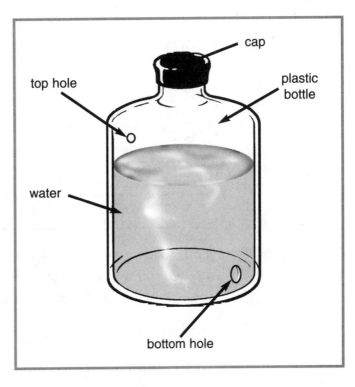

1. Obtain a half gallon (1.9 l) plastic container. Plastic soda bottles will work fine.

2. Place a small hole (about the size of a small nail) at the top as shown.

3. Place a larger hole (about the size of a large nail) at the bottom on the *opposite* end.

4. Do the rest of the experiment outdoors or over a sink.

5. Fill the jar three-fourths full of water. Quickly screw on the cap and place your thumb over the small upper hole.

What happened when you covered the top hole? _____

_____

Say a magic word and release, and then press the hole.

What happens as you uncover the top hole? _____

_____

What must go into the top hole to force water out of the bottom hole? _____

_____

Name _____

Marshmallows and Air Pressure

air out

½ gallon glass jar

large marshmallows

If possible, set up the experiment shown to the left. You will need a strong *glass* jar (about a half-gallon [1.9 l] size) some large marshmallows and a pump that *removes* air.

If you can't set this experiment up, you can still hypothesize about it.

What is mainly inside a marshmallow? _____

_____

What do you think would happen to the air inside the marshmallow if you used the pump to draw

the normal air out of the jar? _____

_____

_____

If possible, try the experiment to see what really happens. Consider an astronaut in a space suit walking on the moon. The moon has no air and therefore no air pressure. What might happen to our moon astronaut if something punctured his suit

and let the air escape? _____

_____

_____

Materials Needed: Glass or straw pipets, jars of water, 50-milliliter graduate, large plastic bag and a half gallon (1.9 l) plastic jar.

For the Teacher: If you can obtain an exhaust pump, set up the marshmallow experiment. The marshmallows will expand greatly as the pressure is reduced.

Glass pipets (unmarked) are best, but straws can be used. You may want to dye the water to increase visibility. Mark a red line with nail polish about 6" (15 cm) from the bottom of the pipet. Encourage sanitary procedures.

52

Name _____

Yeast Under Control

The Scientists' Method

A big part of a scientist's job is to solve problems. In using the scientific method, the scientist formulates hypotheses (educated guesses) and then tests them. This testing often takes the form of a controlled experiment in which everything is kept the same except for one factor that is varied. For example, you might grow one plant in sunlight and another similar plant inside under a bright light. The sunlit plant would be the control. The indoor plant would be the experiment. Everything else besides the light must be exactly the same.

The Science of Yeast

Yeast is a one-celled plant belonging to the order called fungi. It has no chlorophyll and cannot make its own food as green plants do. Yeast cells can break down sugars and starches into carbon dioxide gas (CO_2) and alcohol. Man uses these yeast cells to cause dough to rise. If conditions are right, yeast cells reproduce rapidly by a process called budding. A tiny bud forms on the side of the yeast cell, grows and splits off to become an independent yeast cell.

Bakers must provide the right kind of food for yeasts. Without the right kind of food, they won't produce the gas that causes bread to rise.

Feeding Time for Yeast Cells

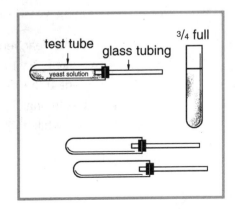

1. Fill two test tubes three-fourths full of warm water.

2. Add half of a level teaspoon (5 ml) of dry yeast to both tubes.

3. Add half of a level teaspoon (5 ml) of sugar to only *one* tube. This will be your *experiment*. Don't put any sugar in the other test tube. This will be your *control*.

Name _____

4. Place a rubber band around the tube with the sugar.

5. Use sticks to stir the ingredients in both test tubes. Use a different stick for each.

6. Place the stoppers with the glass tubing firmly into both test tubes.

7. Place both tubes flat on a table.

You are now waiting for the results of your test. If the sugar can be used as yeast food, gas will form. The gas will push the liquid up the long tube.

Some yeast you buy already has food included in the package. However, if sugar is a good food, the tube with the sugar will bubble up faster.

While you are waiting, describe what you see happening in both tubes. _____

_____

Which tube pushed the fluid up faster? _____

_____

How does this experiment prove that sugar is a good food for yeast? _____

_____

Take the stopper out of the test tube before the fluid overflows. Don't throw anything away.

Place your ear next to the open sugar test tube. Describe what you hear.

_____

_____

Place your nose next to the open sugar test tube. Describe what you smell.

_____

_____

Follow your teacher's directions and **clean up**!

Materials Needed: You will need test tubes, one-hole stoppers, about 6" (15 cm) of glass tubing, yeast, sugar, warm water, rubber bands, teaspoons and stirring sticks.

For the Teacher: For safety reasons, fire polish and insert the glass tubing into the stoppers yourself. Try to find yeast powder with little or no sugar added. Start the controlled experiment at the beginning of the class as it may take 15 plus minutes for the fluid to rise in the glass tubing. Perhaps you can set up some microscopes so your students can observe yeast buds while they wait.

Name _____

"Resilasticity" Lab or How High Will I Bounce?

Why Do Balls Bounce?

All objects in your surroundings differ in their physical characteristics. This lab deals with the properties of *resiliency* and *elasticity* which involve an object being able to return to its original form when compressed. Imagine a rubber ball and yourself being tossed out of a 20-story window. Both you and the ball would be deformed on impact with the sidewalk, but the ball is elastic and resilient enough to resist being deformed, bounce back and restore to its former shape. Let's test a few objects to check their "resilasticity."

Collecting Bouncing Data

Obtain a meterstick, masking tape, graph paper, one Ping-Pong™ ball, one golf ball and one ball of your choice.

1. Tape the meterstick to a table, chair or wall so the 100-centimeter end is up.

2. Hold the Ping-Pong™ ball so that its **bottom** is level with the 100-centimeter mark.

3. Drop the ball and read the height of bounce to the nearest centimeter. Again use the bottom of the ball.

Bring your eyes down to the level of the bouncing ball for the most accurate readings.

4. Record the bounce height in the Ping-Pong™ Data Table under *100-cm DROP.*

5. Repeat two more times and record.

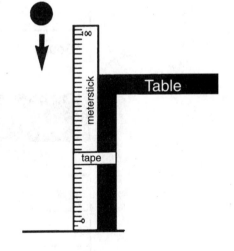

Name _____

6. Average the three readings. Record the average **rounded** off to the nearest whole number. To average, simply add them up and divide by three.

7. Repeat the three trials and averages for 80 to 60 centimeters.

8. Repeat for your Golf Ball Data Table.

9. Repeat for your Ball of Choice Data Table.

PING-PONG™ BALL DATA TABLE
(ALL NUMBERS ROUNDED OFF TO NEAREST CENTIMETER)

TRIAL	100-CM DROP	80-CM DROP	60-CM DROP
1			
2			
3			
Bounce Average			

GOLF BALL DATA TABLE
(ALL NUMBERS ROUNDED OFF TO NEAREST CENTIMETER)

TRIAL	100-CM DROP	80-CM DROP	60-CM DROP
1			
2			
3			
Bounce Average			

BALL OF CHOICE DATA TABLE
(ALL NUMBERS ROUNDED OFF TO NEAREST CENTIMETER)

TRIAL	100-CM DROP	80-CM DROP	60-CM DROP
1			
2			
3			
Bounce Average			

Name _____

Graphing Your Data

A graph is the best way to organize your data. At a glance it can show your data points and enable you to predict what would happen at various heights. The graph your team makes must be: 1) neat, 2) full page, 3) completely labeled, 4) colorful. Follow these instructions using the sample graph shown.

1. Draw a vertical and horizontal line at the edges of your graph paper.

2. Mark off the centimeters for height of drop on the vertical line. Label it *Height of drop in cm.*

3. Mark off the centimeters for height of bounce on the horizontal line. Label it *Height of bounce in cm.*

4. Record your data points for the Ping-Pong™ ball with a big *X* or a heavy dot. Use **averages** only.

5. Draw a line between each data point.

6. Repeat in a **different color** for the other two balls.

7. Put the title *Resilasticity Graph* on top and staple your graph to this lab.

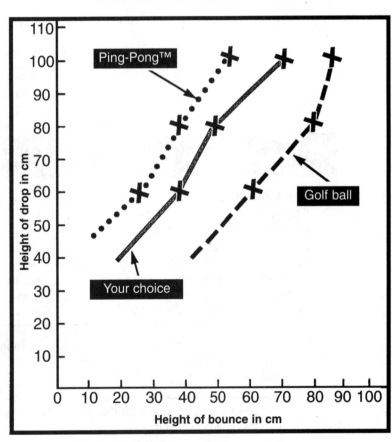

SAMPLE RESILASTICITY GRAPH

Materials Needed: You will need metersticks, tape, graphing or regular paper, Ping-Pong™ balls, golf balls, variety of other balls, crayons or felt pens and rulers.

For the Teacher: This entire lab can be converted to inches if convenient. The optional ball can be provided by you or brought in by the students. "Super Balls" give great results. Your students may need special help graphing. A quick board lesson on graphing would help. If your students can handle it, add a 40-centimeter drop to the data tables. This extra data point makes for a better graph.

Name _____

Inventors: People Who Apply Science

How Well Do You Know the Great American Inventors?

An invention is the creation of a new device, process or product. Inventions help people live better, easier and happier lives. An inventor must have background knowledge, technical ability, creativity and, in many cases, a bit of luck.

Below is a list of only a few of the great American inventors and their contributions. Try to unscramble as many as you can, and write the correct name on the line. Your teacher will provide library research time to help you with the difficult inventors.

INVENTIONS	SCRAMBLED INVENTORS
1. Revolutionary leader who invented the lighting rod and bifocal eyeglasses. 2. American President who invented a new kind of farm plow. 3. Invented the modern form of steamboat.	KANRFNIL _____ FERJSONEF _____ LUTNOF _____
4. Invented the cotton gin that revolutionized agriculture in the South. 5. Invented the reaper that allowed a single farmer to do the work of many. 6. Invented the safety pin. This invention was done in only three hours of work and was sold for only $400.	THWINEY _____ MORCCKICM _____ THUN _____
7. Invented the process of treating rubber to make it more useful. 8. Invented the telegraph. 9. Invented the telephone.	ROYGOAED _____ SMORE _____ LEBL _____
10. He developed over 300 useful products from peanuts to help poor farmers. 11. Invented the sewing machine. 12. Greatest American inventor with over 1000 patents including the incandescent light, the phonograph and the motion picture.	RAVERC _____ WOHE_____ SIDEON _____
13. Invented mass production techniques for manufacturing automobiles. 14. Invented a camera and film system that popularized photography. 15. Invented the Polaroid™ camera for self-processing of film.	DORF _____ MASTANE _____ NADL _____
16. Brothers who developed and flew the first successful airplane. 17. One of the inventors of the transistor. 18. One of the inventors of the laser.	GTHRIW _____ LOCKESYH_____ SOTWEN_____

Name _____

What the World Needs Most

The inventors of the last few years have changed your life in many ways. Computers, robots and biogenetics are making life easier and healthier.

It's time for you to brainstorm a new invention that the world could use. Dream up an invention that would be valuable in factories, homes or "?" Mentally invent something new. Sketch how it would look, and explain your invention's value to mankind.

Inventor's Research

Pick your favorite invention. Prepare a report on the inventor. Give his or her background, and tell how the invention was developed. Bonus points will be given for an oral report with sketches or models.

Materials Needed: You will need either a set of reference books or a trip to the school library.

Name _____

Can You Learn to Think Metric?

The World Is Metric

Practically all scientists use the metric system to measure. Practically all the countries of the world use the metric system. Only the United States and a few small countries still use pounds, quarts and inches. During the Olympics, all events are measured in metric.

Learning the metric system can be fun. It can also be very useful.

The pictures below all deal with metric length. Metric uses centimeters instead of inches, meters instead of yards and kilometers instead of miles. Here are some length comparisons.

1 inch is about 2½ centimeters   1 yard is slightly smaller than a meter

1 mile is slightly over 1½ kilometers

Making Metric Meaningful

Study the pictures below. They will help you estimate metric lengths.

Length 1 cm

= width of paper clip

Length 30 cm

= typical ruler

Length 50 cm

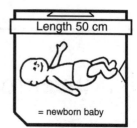

= newborn baby

Length 75 cm

= baseball bat

Length 90 m

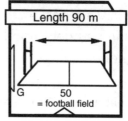

G   50
= football field

Length 5000 km

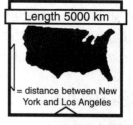

= distance between New York and Los Angeles

Name _____

Guessing in Metric

You will be guessing the metric length of many common objects. Use the metric pictures on page 60 to help you estimate. Fill in the Metric Data Table. First estimate all lengths, then see how close you were by using metric rulers and metersticks.

METRIC DATA TABLE

Objects Measured	Metric Unit	Your Metric Guess	Metric Measurement
Thumbnail width	centimeters		
Thumb length	centimeters		
Arm length (from armpit to fingertip)	centimeters		
Pen width	centimeters		
Pen length	centimeters		
Width of this paper	centimeters		
Length of this paper	centimeters		
Width of your shoe	centimeters		
Length of your shoe	centimeters		
Thickness of a penny	centimeters		
Diameter of a penny	centimeters		
Thickness of a textbook	centimeters		
Width of a textbook	centimeters		
Length of a textbook	centimeters		
Width of your room	meters		
Length of your room	meters		
Height of your room	meters		

Materials Needed: You will need metric rulers and metersticks.

For the Teacher: This activity assumes your students can measure metric length. They may need some help. Stick with centimeters; don't use milliliters. If necessary for your students, let them round off their measured answers to the *nearest* centimeter. Give each team a standard paper clip as an estimation tool.

Name _____

Metric Measurement

Measurement History

One Inch

The United States uses inches, feet and yards to measure length. These units were originally based upon the length of certain body parts.

An inch was the width of a man's thumb.

A foot was the length of a man's foot.

A yard was the distance from a man's nose to the tip of his outstretched arm.

There are obvious problems with using body parts.

What is wrong with using a thumb to measure an inch? _____

_____

What is wrong with using a teacher's foot as the standard foot? _____

_____

One Foot

In this book, you are important. Let's use your thumb width to measure objects.

How many of your thumbs can fit across the width of this page? _____

How many of your thumbs could fit across the length of this page? _____

How many thumb widths are there in the length of your desk? _____

How many thumb widths are there in the length of your shoe? _____

One Yard

List some problems using your thumb, or any one's, as a measuring unit.

_____

_____

Down with Inches, Feet and Yards

Only the United States and a few other countries use the inches, feet and yards measurement system. Most countries in the world and most scientists use the **metric system**. Metric is easier to use once you learn it.

Name _____

Metric length is not based on body parts. It is based on the distance between the equator and the North Pole. The metric **meter** is $1/10,000,000$ of that distance. A meter is **about** 4" (10 cm) longer than one yard. A meter is divided into 100 parts called **centimeters**. Study the metric ruler below. It is designed to help you read centimeters. *It is not exact.*

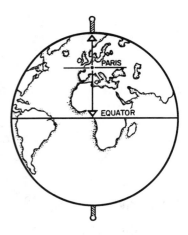

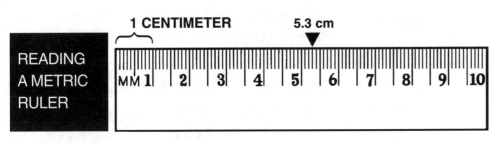

READING A METRIC RULER

Below is an *exact* metric ruler.

1 centimeter   5.3 cm

1   2   3   4   5   6   7   8   9   10

A   B   C   D   E

How many centimeters are there at each arrow?

Arrow A _____ centimeters   Arrow B _____ centimeters   Arrow C _____ centimeters

Arrow D _____ centimeters   Arrow E _____ centimeters

Metric Practice

Obtain a metric ruler. Measure the distances below to the nearest tenth of a centimeter.

What is the width of this page? _____ centimeters

What is the length of this page? _____ centimeters

What is the length of a standard paper clip? _____ centimeters

What is the width of a standard paper clip? _____ centimeters

What is the diameter (longest distance across) of a penny? _____ centimeters

What is the diameter of a nickel? _____ centimeters

What is the length of your pen? _____ centimeters

What is the width of your thumb? _____ centimeters

What is the length of your shoe? _____ centimeters

Materials Needed: Metric rulers or metersticks.

Answer Key

Observation Tests, page 8

Observation Test 1: Scissors won't close

Observation Test 2: *The* is repeated on the sign.

Observation Test 3: 16 Fs

Observation Test 4: Hilda can't be a widow; she died first.

Latin Scientists, page 18

1. flight and space
2. body construction
3. primitive man
4. ancient civilizations
5. skin
6. insects
7. races
8. plants
9. birds
10. poisons
11. weather
12. fossils
13. body functions
14. blood
15. fish

The Cool Tools of Science, page 19

flasks–2	balance–11
thermometer–12	test tube–4
microscope–10	graduate–3
Petri dish–5	tweezers–7
beaker–1	ring stand and clamp–6
burner–8	

First Scientific Problem, page 22

Normally a weight at the bottom does balance best. However, in this situation, you're balancing it with your finger. The weighted stick on top takes longer to fall, and you have more time to get it back in balance.

Second Scientific Problem, page 23

The ball on the pie pan follows Newton's first law and continues on in a straight line as shown by path B.

Third Scientific Problem, pages 23-24

Assuming the balloons are new, the small balloon should blow up the large balloon. This is because the rubber of the small balloon still has lots of elasticity. The greatly expanded balloon has lost its elasticity. Refer to their experience in having initial difficulty blowing up a balloon. As the balloon gets larger, it is easier to blow up.

How Well Do You Know the Great American Inventors? page 58

1. Franklin	2. Jefferson
3. Fulton	4. Whitney
5. McCormick	6. Hunt
7. Goodyear	8. Morse
9. Bell	10. Carver
11. Howe	12. Edison
13. Ford	14. Eastman
15. Land	16. Wright
17. Shockley	18. Townes